4

Boozy Shakes

Boozy Shakes

Milkshakes, malts and floats for grown-ups

Victoria Glass

Photography by Gareth Morgans

RYLAND PETERS & SMALL

LONDON • NEW YORK

Designer Maria Lee Warren
Commissioning Editor
Stephanie Milner
Head of Production Patricia Harrington
Art Director Leslie Harrington
Editorial Director Julia Charles
Publisher Cindy Richards

Styling and Photographic Art Direction Luis Peral
Food Stylists Lizzie Harris and Kim Morphew
Indexer Vanessa Bird

First published in 2015
by Ryland Peters & Small
20–21 Jockey's Fields,
London WC1R 4BW
and
341 E 116th St
New York, NY 10029
www.rylandpeters.com

10 9 8 7 6 5 4 3 2 1

ISBN: 978-1-84975-608-2

Printed and bound in China

A CIP record for this book is available from the British Library.

US Library of Congress Cataloging-in-Publication Data has been applied for.

Note

- Both British (Metric) and American (Imperial plus US cups) are included in these recipes for your convenience, however it is important to work with one set of measurements and not alternate between the two within a recipe.
- All spoon measurements are level unless otherwise specified.
- When a recipe calls for the grated zest of citrus fruit, buy unwaxed fruit and wash well before using. If you can only find treated fruit, scrub well in warm soapy water before using.
- Cream goes off quickly, so always practise food safety by using fresh cream before its expiry date.
- These recipes are not intended for consumption by children. If you are over the legal purchase age and choose to drink alcohol, always do so responsibly and in moderation.

Author's acknowledgements
Heartfelt thanks to my wonderful agent, Olivia Guest, for all her support and to the amazing team at Ryland Peters & Small for helping to create such a beautiful book. Huge thanks to Gareth Morgans and his assistant, Alex Luck, for the stunning photography and to Maria Lee Warren for her fabulous design. Thanks also to Luis Peral for his stellar styling and art direction, Patricia Harrington on production and the brilliant Home Economists, Lizzie Harris and Kim Morphew. Special thanks must go to Julia Charles, Cindy Richards and Leslie Harrington, but particularly to my wonderful editor, Stephanie Milner, for her excellent advice and humour. And finally, thanks to the gorgeous Richard Hurst, who dutifully and selflessly knocked back enough test shakes to warrant a lie down afterwards.

Contents

Introduction

Remember the excitement when you were a child, of being handed a giant sundae glass full of creamy milkshake with not one, but two bendy straws? It's time to give nostalgia a grown-up makeover, with a hearty measure of hard liquor added to the mix. Behold the beauty of the hard shake! As an accompaniment to barbecues, pulled pork, or just on their own at cocktail hour, these drinks range from playful to sophisticated.

Milkshakes have long been the perfect pairing with highway café chalkboard offerings of corn dogs, cheeseburgers and fried chicken. As deliciously 'dirty' food has taken its place on gourmet menus, a new style of milkshake is needed to keep up the pace. And what better way to fuel this milkshake evolution, than by unlocking the drinks cabinet? If you want your choice of burger to be more of an event than a trip to the nearest golden arches, then you'll need a drink that's a more suitable pairing than a diet cola.

Cold, sweet and gluggably moreish, boozy shakes combine all the appeal of a trip to the diner with all that glitters behind the bar. No longer must you choose between a bourbon or a chocolate malt to go with your burger: my Harlem Shake contains the best of both worlds. I have created shakes based on favourite confections, cocktails, desserts and even music. In this collection of recipes, you can't fail to find something to get your taste buds dancing, whatever the occasion and whatever your mood.

Take your time over 1950s diner style long shakes with a Three Sheets to the Wind Shamrock Shake or the king of hard shakes, The Elvis. Get nostalgic with a booze-fuelled Piña Colada or a schnapps-y Peach Melba. If it's a cocktail shake you're after, you can do worse than to sip on a sophisticated Aztec Margarita or Hazelnut Martini. If a lighter approach to party time is in order, try the Let England Shake or Dark & Stormy Float, which have all the froth, fun and kick of a hard shake, but won't fill you up too much before dinner. And if you're skipping dessert in favour of a refill, why not go for an alcohol-infused Lemon Meringue Pie or Cookies & Irish Cream.

Boozy shakes might not win any prizes for nutrition, but they'll definitely hit the top of the scoreboard for deliciousness. Besides, the occasional indulgence is good for the soul and a Bananas Foster can even count towards your fruit and vegetable five-a-day. These strictly adult treats are a welcome, frothy addition to any table, especially if you like your drinks with a dollop of indulgence and a super-sized shot from the top shelf.

Stir up a shake and kick back, because milkshakes just got drunk. It's time to say, 'Cheers!'

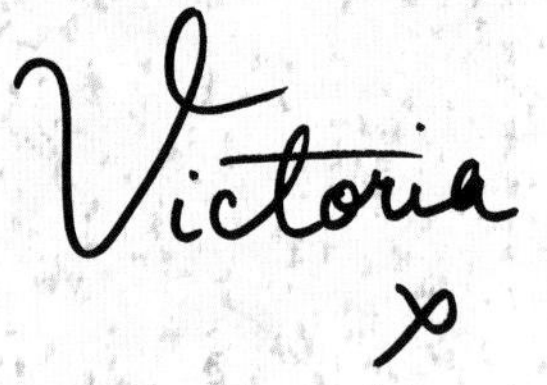

Basic Recipes

Vanilla Ice Cream

350 ml/1 1/3 cups double/heavy cream
1 vanilla pod/bean, scored lengthways and seeds scraped out
4 egg yolks
100 g/1/2 cup caster/granulated sugar

an ice cream maker (optional)

Makes 500 ml/1 pint

Put the cream in a saucepan or pot with the vanilla pod/bean and seeds set over a gentle heat. While the cream heats, whisk together the egg yolks and sugar in a heatproof bowl until pale and creamy. Rest a fine mesh sieve/strainer over the bowl. Once the cream has come to the boil, remove it from the heat and pour through the sieve/strainer set over the egg mixture. Discard the vanilla. Whisk the egg and hot cream mixture together and pour back into the pan. Return to a gentle heat and whisk constantly until the custard thickens enough so that it coats the back of a spoon. Immediately remove from the heat and pour the custard into a jug/pitcher. Cover the top with clingfilm/plastic wrap to prevent a skin forming and leave to cool completely before transferring to the fridge to chill for at least 2 hours. If using an ice cream maker, follow the manufacturer's instructions. If making by hand, simply pour the custard into an airtight container and freeze for about 4 hours or until set, giving the mixture a vigorous whisk every 30 minutes to prevent ice crystals from forming.

Strawberry Sorbet

175 g/3/4 cup plus 2 tablespoons caster/granulated sugar
500 g/5 cups fresh strawberries, hulled
1/4 large cucumber, chopped
2 teaspoons freshly squeezed lemon juice
a handful of fresh mint, finely chopped

an ice cream maker (optional)

Makes 1 litre/2 pints

Begin by making a sugar syrup. Put the sugar and 175 ml/3/4 cup of water in a saucepan or pot set over a gentle heat and stir until the sugar has dissolved. Remove from the heat and cool completely. Put the strawberries, cucumber and lemon juice into the blender with the cold sugar syrup and blitz until liquidized. Pass the mixture through a fine mesh sieve/strainer into a jug/pitcher and chill in the fridge for at least 4 hours. If using an ice cream maker, pour the liquid into the machine and once it begins to set, add the chopped mint and continue to churn until fully set. Transfer the sorbet to an airtight container and store in the freezer until needed. If making by hand, simply stir in the mint straight away and pour the liquid into an airtight container and freeze for about 4 hours, or until set, giving the mixture a vigorous whisk every 30 minutes to prevent crystals forming.

Cherry Sauce

450 g/2½ cups fresh cherries, stoned/pitted
25 g/2 tablespoons caster/granulated sugar
2 teaspoons freshly squeezed lemon juice

Makes 315 ml/1⅓ cups

Put all of the ingredients in a saucepan or pot with 1 tablespoon of water. Set the mixture over a gentle heat and warm through, stirring every now and then, until the sugar has dissolved and the fruit has completely broken down. Remove from the heat and cool completely before blitzing in the blender until smooth.

Chocolate Fudge Sauce

100 g/½ cup light muscovado sugar
50 ml/3 tablespoons double/heavy cream
2 tablespoons golden/light corn syrup
100 g/¾ cup chopped dark/bittersweet chocolate

Makes 250 ml/1 cup

Put all of the ingredients in a saucepan or pot set over a gentle heat and stir until the chocolate has melted and the sugar has dissolved. You should have a thick, glossy sauce. Transfer to a jug and leave to cool before using.

Cinnamon Fudge Sauce

100 g/½ cup light muscovado sugar
3 tablespoons golden/light corn syrup
30 g/2 tablespoons butter
2 teaspoons pure vanilla extract
a pinch of salt
2 teaspoons ground cinnamon
125 ml/½ cup double/heavy cream

Makes 300 ml/1¼ cups

Put all of the ingredients, except for the cream, in a saucepan or pot set over a gentle heat and stir until the sugar has dissolved. Bring to a rolling boil, before stirring in the cream. Reduce the heat and simmer gently for 1–2 minutes. Transfer to a jug/pitcher and leave to cool before using.

Peanut Butter Butterscotch Sauce

100 g/½ cup light muscovado sugar
3 tablespoons golden/light corn syrup
30 g/2 tablespoons butter
2 teaspoons pure vanilla extract
a pinch of salt
2 tablespoons smooth peanut butter
150 ml/⅔ cup double/heavy cream (or condensed milk)

Makes 300 ml/1¼ cups

Follow the instructions for the Cinnamon Fudge Sauce above.

Raspberry Sauce

450 g/4 cups fresh raspberries
25 g/2 tablespoons caster/granulated sugar
2 teaspoons freshly squeezed lemon juice

Makes 315 ml/1⅓ cups

Follow the instructions for the Cherry Sauce above, then push the mixture through a fine mesh sieve/strainer set over a jug/pitcher to remove the seeds and pulp.

Whisky Butterscotch Sauce

100 g/½ cup light muscovado sugar
2 tablespoons golden/light corn syrup
30 g/2 tablespoons butter
2 teaspoons pure vanilla extract
a pinch of salt
3 tablespoons whisky
125 ml/½ cup double/heavy cream (or condensed milk)

Makes 300 ml/1¼ cup

Follow the instructions for the Cinnamon Fudge Sauce opposite.

Blueberry Syrup

150 g/1¼ cups fresh blueberries
50 ml/3 tablespoons golden/light corn syrup

Makes 125 ml/½ cup

Put all of the ingredients in a saucepan or pot set over a gentle heat. Stir every now and then until the blueberries have broken down. Transfer to a jug/pitcher and leave to cool before using.

Coffee Syrup

1 tablespoon instant espresso powder
100 g/½ cup caster/granulated sugar
200 ml/¾ cup water

Makes 150 ml/½ cup plus 2 tablespoons

Put all of the ingredients in a saucepan or pot set over a gentle heat and stir until the coffee and sugar have dissolved. Continue to simmer until the syrup thickens slightly. Transfer to a jug/pitcher and leave to cool before using.

Rhubarb Compote

400 g/4 cups 5-cm/2-in. fresh rhubarb pieces, preferably forced pink rhubarb
110 g/generous ½ cup caster/granulated sugar
finely grated zest of 1 orange
1 vanilla pod/bean, scored lengthways and seeds scraped out

Makes 500 ml/2 cups plus 2 tablespoons

Put the rhubarb and sugar together in a large mixing bowl and toss to coat the rhubarb in sugar. Cover and set aside for 30 minutes. The rhubarb will release some of its juice.

Place the rhubarb and juice in a saucepan or pot with the remaining ingredients and set over a gentle heat. Stir until the sugar has dissolved, then simmer until the rhubarb is soft. Remove it from the heat and pour through the sieve/strainer set over a bowl. Discard the vanilla and leave to cool completely. Cover and store in the fridge for up to 2 weeks.

Swiss Meringue

110 g/generous ½ cup caster/granulated sugar
2 egg whites

an electric handheld whisk

a large mixing bowl, chilled in the fridge

Makes 500 ml/2 cups

Put the sugar and egg whites in a heatproof bowl set over a saucepan or pot of barely simmering water. Ensure that the bowl does not touch the water. Whisk on a medium setting until the sugar has dissolved and the mixture begins to look like whipped cream. Insert a sugar thermometer (or thermapen) and continue to whisk until the meringue reaches 55°C/130°F. Remove from the heat and transfer to a cold bowl until needed.

The Candy Bar

No British childhood would be complete without these rhubarb and custard boiled sweets from the sweet shop, served up in a striped paper bag. This boozy shake is full of nostalgia for adults only.

Rhubarb & Custard

300 ml/10 oz. advocaat
150 ml/5 oz. rhubarb vodka
4 scoops Vanilla Ice Cream (see page 7)
100 ml/⅓ cup whole milk
4 tablespoons Rhubarb Compote (see page 9)

2 highball tumblers

Makes 900 ml/30 oz. and serves 2

Place the glasses in the freezer to chill for a few minutes.

Blend together the advocaat, rhubarb vodka, ice cream, milk and half of the rhubarb compote until smooth and thick.

Place 1 tablespoon of rhubarb compote in the base of each glass and swirl it up the sides. Divide the milkshake between the two glasses and serve.

This is based on the classic flavour pairing of Terry's Chocolate Orange, a confection shaped like an orange that is cracked open and eaten in segments. There's no need to 'tap and unwrap' this boozy shake.

Chocolate Orange

grated zest and freshly squeezed juice of 1 orange
100 ml/3½ oz. Cointreau (or other orange liqueur)
100 ml/3½ oz. crème de cacao (dark)
100 ml/⅓ cup whole milk
4 scoops Chocolate Ice Cream (see below)

Chocolate Ice Cream
1 quantity Vanilla Ice Cream (see page 7)
100 g/¾ cup dark/bittersweet chocolate chips, melted
a pinch of salt

4 lowball fish-bowl glasses

Makes 800 ml/27 oz. and serves 4

Begin by making the Chocolate Ice Cream or use a good-quality store-bought ice cream. Follow the instructions for making Vanilla Ice Cream on page 7, but add the melted chocolate and a small pinch of salt to the hot custard and whisk them in before chilling.

Place the glasses in the freezer to chill for a few minutes.

For the shake, blend together two-thirds of the orange zest, the juice, Cointreau, crème de cacao and ice cream until thick and smooth. Add the milk and blend again to combine.

Divide between fish-bowl glasses and sprinkle with the remaining orange zest to serve.

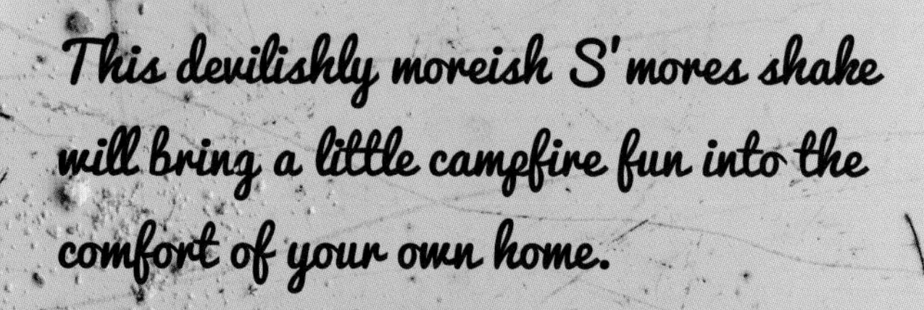
This devilishly moreish S'mores shake will bring a little campfire fun into the comfort of your own home.

Drunken S'mores

75 ml/2½ oz. vanilla vodka
75 ml/2½ oz. crème de cacao (dark)
4 scoops Chocolate Ice Cream (see below)

Chocolate Ice Cream
1 quantity Vanilla Ice Cream (see page 7)
100 g/¾ cup dark/bittersweet chocolate chips, melted
a pinch of salt

To serve
2 tablespoons Chocolate Fudge Sauce (see page 8)
canned whipped cream
50 g/½ cup digestive biscuit/graham cracker crumbs
a handful of mini marshmallows

2 milkshake glasses

Makes 600 ml/21 oz. and serves 2

Begin by making the Chocolate Ice Cream or use a good-quality store-bought ice cream. Follow the instructions for making Vanilla Ice Cream on page 7, but add the melted chocolate and a small pinch of salt to the hot custard and whisk them in before chilling.

Place the glasses in the freezer to chill for a few minutes.

For the shake, blend together the alcohol and ice cream until smooth and thick. Drop a tablespoon of Chocolate Fudge Sauce into each glass and divide the milkshake between them.

Squirt a generous swirl of cream on top of each milkshake. Sprinkle with crumbs and tumble on as many mini marshmallows as you can fit. Toast the tops of the marshmallows with a chef's blowtorch before serving, if you wish.

Peanut butter cups are so good, they're nigh on impossible to improve upon. But that doesn't mean you shouldn't try. I've given it my best shot with this bourbon and peanut butterscotch-laced libation.

Peanut Butter Cup

4 scoops Peanut Butter Ice Cream (see below)
50 ml/1¾ oz. bourbon
50 ml/1¾ oz. whole milk
1 tablespoons crème de cacao (brown)
2 tablespoons Chocolate Fudge Sauce (see page 8)
2 tablespoons Peanut Butter Butterscotch Sauce (see page 8)

Peanut Butter Ice Cream
1 quantity Vanilla Ice Cream (see page 7)
125 g/½ cup smooth peanut butter
a pinch of salt
10 peanut butter cups, such as Reese's

To serve
canned whipped cream
peanut butter cups, chopped

2 coupe à glace glasses

Makes 375 ml/13 oz. and serves 2

Begin by making the Peanut Butter Ice Cream or use a good-quality store-bought ice cream. Follow the instructions for making Vanilla Ice Cream on page 7, but add the peanut butter and a small pinch of salt to the hot custard and whisk in thoroughly before chilling. Leave to cool completely before mixing in the chopped-up peanut butter cups before freezing or churning in the ice cream maker.

Place the glasses in the freezer to chill for a few minutes.

For the shake, put the ice cream, bourbon, milk, crème de cacao and 1 tablespoon each of Chocolate Fudge Sauce and Peanut Butter Butterscotch Sauce in the blender and pulse until smooth and thick.

Swirl the remaining 1 tablespoon of Chocolate Fudge Sauce around the base and insides of the glasses. Divide the milkshake between the glasses and top with canned whipped cream. Drizzle the remaining 1 tablespoon of Peanut Butter Butterscotch Sauce over the cream and sprinkle over the chopped peanut butter cups before serving.

Aromatic and heady with Ottoman flavours, Turkish delight is an ancient treat that has stood the test of time for a reason. This shake is pure indulgence.

Turkish Delight

5 scoops Pistachio Ice Cream (see below)
75 ml/2½ oz. vanilla vodka
1–2 teaspoons rose water, to taste
45 ml/3 tablespoons whole milk
100 g/4 oz. (about 8 pieces) Turkish delight, finely chopped
8 teaspoons Grenadine
crystallized rose petals

Pistachio Ice Cream
1 quantity Vanilla Ice Cream (see page 7)
125 g/⅓ cup unsweetened pistachio purée/paste
a pinch of salt

4 coupe glasses

Makes 500 ml/17 oz. and serves 4

Begin by making the Pistachio Ice Cream or use a good-quality store-bought ice cream. Follow the instructions for making Vanilla Ice Cream on page 7, but whisk the pistachio purée/paste and a small pinch of salt into the hot custard before the chilling stage.

Place the glasses in the freezer to chill for a few minutes.

For the shake, put the ice cream, vanilla vodka, 1 teaspoon rose water and milk in a blender and pulse until combined. Taste for rose, adding more to taste if needed. Mix in the finely chopped Turkish delight.

Splash the inside of each glass with 2 teaspoons of Grenadine. Divide the milkshake among the glasses and top each with crystallized rose petals to serve.

Milk break has never been so much fun with this Amaretto-spiked Marzipan Milk. It may look innocent, but it's definitely best kept away from the school yard.

Marzipan Milk

Place the bottles in the freezer to chill for a few minutes.

Put all the ingredients in a blender and pulse until smooth and frothy.

Divide the milkshake between the mini milk bottles using a funnel, and serve with stripy straws.

150 ml/5 oz. Amaretto di Saronno® (or other almond liqueur)
6 scoops Vanilla Ice Cream (see page 7)
250 ml/1 cup whole milk
a few drops of pure almond extract, to taste

4 mini milk bottles

Makes 800 ml/28 oz. and serves 4

The Cake Shop

Chocolate, cherries and kirsch – what's not to like? It's the 1970s classic, Black Forest Gâteau, sucked through a straw.

Black Forest Frappé

100 ml/3½ oz. kirsch
2 scoops Chocolate Ice Cream (see below)
3 ice cubes
75 ml/¼ cup whole milk
5 tablespoons Cherry Sauce (see page 8)

Chocolate Ice Cream
1 quantity Vanilla Ice Cream (see page 7)
100 g/¾ cup dark/bittersweet chocolate chips, melted
a pinch of salt

To serve
50 g/scant ½ cup (about 12) kirsch-soaked cherries, finely chopped
canned whipped cream
chocolate shavings
3 fresh cherries

2 milkshake glasses

Makes 500 ml/17 oz. and serves 2

Begin by making the Chocolate Ice Cream or use a good-quality store-bought ice cream. Follow the instructions for making Vanilla Ice Cream on page 7, but add the melted chocolate and a small pinch of salt to the hot custard and whisk them in before chilling.

Place the glasses in the freezer to chill for a few minutes.

For the shake, blend together the kirsch, ice cream, ice cubes and milk until smooth and thick. Add in 2 tablespoons of the Cherry Sauce. Stir through the finely chopped kirsch-soaked cherries.

Add 1 tablespoon of Cherry Sauce to the bottom of each glass and pour the milkshake over it. Top each milkshake with a generous cloud of whipped cream and scatter over some chocolate shavings. Put a fresh cherry on top and enjoy!

Trifle with raspberries and Scotch whisky liqueur, or straight-up whisky if you prefer, mashed into a milkshake. There are harder ways to spend your time than drinking this winning combination.

Tipsy Laird

Place the glasses in the freezer to chill for a few minutes.

Blend together the cake crumbs, ice cream, Drambuie or whisky liqueur and milk.

Place a tablespoon of raspberry sauce in the bottom of each glass and pour the milkshake over the top. Top with a generous squirt of cream, sprinkle over some toasted almond flakes and tumble on some fresh raspberries.

50 g/½ cup vanilla cake crumbs
4 scoops Vanilla Ice Cream (see page 7)
100 ml/3½ oz. Drambuie (or whisky liqueur)
100 ml/3½ oz. whole milk
4 tablespoons Raspberry Sauce (see page 8)
canned whipped cream
10–12 fresh raspberries
flaked/slivered almonds, toasted

4 coupe à glace glasses

Makes 650 ml/22 oz. and serves 4

I've turbo-charged a classic cookies-and-cream shake with a generous measure of Irish cream liqueur, making this sweet treat strictly for grown-ups.

Cookies & Irish Cream

Place the glasses in the freezer to chill for a few minutes.

Put all of the ingredients (except for the cookies used for garnishing the glasses) in a blender and pulse until smooth and thick.

Divide the milkshake between the chilled glasses and edge each glass with a cookie – scrape some of the filling out of one half of each cookie and hook over the side of the glass like a lime wedge.

200 ml/7 oz. Irish cream liqueur, such as Baileys®
150 ml/$^2/_3$ cup whole milk
8 scoops Vanilla Ice Cream (see page 7)
8 chocolate sandwich cookies, such as Oreo, plus 2, to serve

2 milkshake glasses

Makes 800 ml/26 oz. and serves 2

For full authenticity, use Key limes, which have a higher acidity and are more aromatic than green Persian limes, though these will do the job nicely if that's all you can find.

Key Lime Pie

grated zest and freshly squeezed juice of 2 limes, plus extra zest to garnish
100 ml/3½ oz. vodka (or lime vodka)
4 scoops Lime and Condensed Milk Ice Cream (see below)
100 ml/⅓ cup whole milk
1 quantity Swiss Meringue (see page 9)
50 g/½ cup digestive biscuit/Graham cracker crumbs

Lime Ice Cream
200 ml/¾ cup condensed milk
250 ml/1 cup double/heavy cream
finely grated zest and freshly squeezed juice of 2 limes

2 lowball tumblers

a piping/pastry bag fitted with a star nozzle/tip

Makes 600 ml/20 oz. and serves 2

Begin by preparing the Lime Ice Cream. Whisk all the ingredients together until very thick and transfer to a airtight plastic container. Simply place it in the freezer for at least 6 hours – there's no need to churn it.

Place the glasses in the freezer to chill for a few minutes.

For the shake, blend together the lime zest and juice, vodka and ice cream until thick and smooth. Add the milk and blend again.

Divide the milkshake between the glasses. Fill the piping/pastry bag with the Swiss Meringue. Pipe a generous swirl of meringue on top of each milkshake. Sprinkle with the crushed biscuit/cracker crumbs and a little extra lime zest before serving.

This classic dessert pie boasts fresh and zesty lemon curd topped with baked meringue. My version has a zingy hit of Limoncello and a few scoops of tangy ice cream to shake this pie up to new milky heights.

Lemon Meringue Pie

grated zest and freshly squeezed juice of 2 lemons
100 ml/3½ oz. Limoncello
4 tablespoons lemon curd
8 scoops Lemon Curd Ice Cream (see below)
4 mini meringue nests

Lemon Curd Ice Cream
1 quantity Vanilla Ice Cream made without the vanilla pod/bean (see page 7)
3–4 tablespoons lemon curd

a handheld electric blender or milkshake mixer

3 lowball tumblers

Makes 600 ml/20 oz. and serves 3

Begin by preparing the Lemon Curd Ice Cream or use a good-quality store-bought ice cream. Follow the instructions for making Vanilla Ice Cream on page 7, omitting the vanilla pod/bean. Once the custard has been made, stir in 2 tablespoons of lemon curd and leave to chill. When the ice cream is churning in the ice cream maker and begins to set, add another 1–2 tablespoons of lemon curd to the ice cream for a rippled effect.

Place the glasses in the freezer to chill for a few minutes.

For the shake, put the lemon zest and juice, limoncello, 2 tablespoons of the lemon curd and the ice cream into a jug/pitcher. Pulse lightly with a handheld electric blender, then switch to whisking by hand, to help keep your shake nice and thick. If you have a milkshake mixer, blend it in the cup of the mixer until smooth and thick.

Swirl the remaining lemon curd around the edges and bottoms of the glasses. Pour in the milkshake before crushing the meringue nests over the top to serve.

This retro dessert has been given a modern makeover in this boozy banana milkshake. Your kids will definitely want to muscle in on the fun, but these shakes are strictly for adults only.

Bananas Foster

100 ml/3½ oz. dark rum
30 ml/1 oz. crème de banane
2 large bananas, peeled and roughly chopped
5 scoops Vanilla Ice Cream (see page 7)
4 tablespoons Cinnamon Fudge Sauce (see page 8)
200 ml/¾ cup whole milk
ground cinnamon, to dust

2 hurricane glasses

Makes 850 ml/29 oz. and serves 2

Place the glasses in the freezer to chill for a few minutes.

Blend together the dark rum, crème de banane, bananas, Vanilla Ice Cream, 2 tablespoons of the Cinnamon Fudge Sauce and milk until thick and smooth.

Place 1 tablespoon of Cinnamon Fudge Sauce in the base of each glass and swirl it up the sides. Divide the milkshake between the two glasses and dust their tops with ground cinnamon.

The Peach Melba was invented to honour the Australian soprano, Nellie Melba. This quaffable version with added peach schnapps will certainly get your tastebuds singing.

Peach Melba

1 x 420-g/16-oz. can peach slices in light syrup
200 ml/7 oz. peach schnapps
4 scoops Vanilla Ice Cream (see page 7)
6 tablespoons Raspberry Sauce (see page 8)

a handheld electric blender

2 milkshake glasses

Makes 800 ml/27 oz. and serves 2

Place the glasses in the freezer to chill for a few minutes.

Drain most of the syrup from the peaches and use a handheld electric blender to purée the fruit. Set aside.

Blend together the peach schnapps and Vanilla Ice Cream until smooth and thick. Add the peach purée and stir to combine. Swirl 3 tablespoons of the Raspberry Sauce around the insides of each glass. Pour over the peach milkshake at an angle to create an attractive two-tone pattern in the milkshake and serve.

The Cocktail Shaker

Vodka, Baileys, Kahlúa, milk and ice cream: I defy you not to climb aboard this Mudslide Shake train.

Mudslide

Place the glasses in the freezer to chill for a few minutes.

Put the vodka, Kahlúa, Baileys, milk and ice cream in a blender and pulse until smooth.

Pour a tablespoon of Coffee Syrup into the base of each glass and swirl it up the insides of each glass. Divide the milkshake between the glasses. Top with a swirl of canned whipped cream and decorate with a few chocolate-covered coffee beans.

75 ml/2½ oz. vodka
75 ml/2½ oz. Kahlúa
75 ml/2½ oz. Baileys
100 ml/⅓ cup whole milk
2 scoops Vanilla Ice Cream (see page 7)
2 tablespoons Coffee Syrup (see page 9)
canned whipped cream
chocolate-covered coffee beans (optional)

2 lowball tumblers

Makes 400 ml/14 oz. and serves 2

With two parts alcohol to one part cream, a Brandy Alexander is practically a hard shake already. With the addition of milk and lashings of vanilla ice cream, my recipe turns the already rich cocktail into a long dessert shake.

Brandy Alexander

60 ml/2 oz. cognac
60 ml/2 oz. crème de cacao (brown)
60 ml/¼ cup double/heavy cream
60 ml/¼ cup whole milk
4 scoops Vanilla Ice Cream (see page 7)

To garnish
canned whipped cream
dark/bittersweet chocolate, for grating
nutmeg, for grating

4 martini glasses

Makes 450 ml/15 oz. and serves 4

Place the glasses in the freezer to chill for a few minutes.

Put the cognac, crème de cacao, cream, milk and ice cream in a blender and pulse until smooth and thick.

Divide the milkshake between the chilled martini glasses and top with canned whipped cream and a little freshly grated chocolate and nutmeg.

This sophisticated shake, makes the most of chocolate and spice – a classic flavour combination since pre-Columbian Mesoamerica.

Aztec Margarita

Begin by making the Chocolate Ice Cream, following the instructions on page 13, or use a good-quality store-bought ice cream.

Place the glasses in the freezer to chill for a few minutes.

For the shake, lightly crush the cardamom pods using a pestle and mortar, and break up the cinnamon stick. Put the milk and cream in a small saucepan, add the spices and simmer until the cream begins to boil. Remove the pan from the heat and set aside until cold, to allow the spices to infuse with the milk and cream. Pass the cold cream through a fine mesh sieve/strainer set over a bowl. Discard the spices and chill the cream until needed.

Next, decorate the glasses. Mix together the sugar and cinnamon and place in a shallow bowl wide enough to fit the margarita glasses. Rub a lime wedge around the edge of each chilled glass before upturning them into the cinnamon sugar. Twist to coat the edge, then set aside.

To make the shake, put the chilled infused cream, tequila, vodka, triple sec, Chocolate Ice Cream and Chocolate Fudge Sauce in a blender with 2 drops of Tabasco. Pulse until smooth and frothy. Taste and add more Tabasco if desired. Blitz again and pour into the prepared glasses.

6 scoops Chocolate Ice Cream (see page 13)
15 cardamom pods
1 cinnamon stick
45 ml/3 tablespoons whole milk
45 ml/3 tablespoons double/heavy cream
120 ml/4 oz. gold tequila
60 ml/2 oz. chilli/hot pepper vodka
60 ml/2 oz. triple sec
3 tablespoons Chocolate Fudge Sauce (see page 8)
3–4 drops of Tabasco

Cinnamon Sugar-rimmed Glasses
25 g/2 tablespoons caster/granulated sugar
1 teaspoon ground cinnamon
½ lime, cut into wedges

4 margarita glasses

Makes 800 ml/28 oz. and serves 4

A Whisky Mac is the perfect winter warmer, but my reinvention can take you through to the summer months thanks to its indulgent addition of aromatic Ginger Ice Cream.

Whisky Mac

Begin by making the Ginger Ice Cream or use a good-quality store-bought ice cream. Follow the instructions for making Vanilla Ice Cream on page 7, omitting the vanilla pod/bean and whisking the ground ginger into the cream in the saucepan instead. Add the chopped stem ginger and ginger syrup to the custard before chilling.

Place the glasses in the freezer to chill for a few minutes.

For the shake, put the ice cream, whisky, ginger wine and milk in a blender and pulse until smooth and frothy. Pour a tablespoon of the Whisky Butterscotch Sauce in the base of each glass and swirl it up the sides. Divide the milkshake between the glasses and serve.

4 scoops Ginger Ice Cream (see below)
55 ml/1½ oz. Scotch whisky
30 ml/1 oz. ginger wine
65 ml/¼ cup whole milk
2 tablespoons Whisky Butterscotch Sauce (see page 9)

Ginger Ice Cream
1 quantity Vanilla Ice Cream made without the vanilla pod/bean (see page 7)
2 teaspoons ground ginger
2 balls stem ginger in syrup, finely chopped
4 tablespoons syrup from the jar of stem ginger

2 highball tumblers

Makes 400 ml/14 oz. and serves 2

Almond liqueur and cherries with an added zing of lemon make for a seriously sophisticated and sexy shake.

Amaretto Sour

150 g/1 cup fresh cherries, stoned/pitted
finely grated zest and freshly squeezed juice of 1 lemon
2 tablespoons maraschino liqueur
80 ml/2½ oz. Amaretto di Saronno®
2 scoops Vanilla Ice Cream (see page 7)
100 ml/⅓ cup whole milk
canned whipped cream
4 maraschino cocktail cherries

4 lowball tumblers

Makes 600 ml/20 oz. and serves 4

Place the glasses in the freezer to chill for a few minutes.

Begin by making cherry purée: blend the fresh cherries in a food processor then pass them through a fine mesh sieve/strainer. You should have about 100 ml/⅓ cup.

Blend together the cherry purée, lemon zest and juice and the alcohol until fully mixed. Add the ice cream and blend until smooth. Add the milk and blend one last time, before dividing between the glasses. Top with a modest squirt of cream and a maraschino cherry.

A Dark & Stormy is the perfect summer cocktail. With an added few scoops of Ginger and Lime Ice Cream, summer just got cooler.

Dark & Stormy Float

2–3 scoops Ginger and Lime Ice Cream (see below)
50 ml/1¾ oz. dark rum
275 ml/1 cup and 1 tablespoon fiery and aromatic ginger beer, such as Fentimans

Ginger and Lime Ice Cream
1 quantity Vanilla Ice Cream made without the vanilla pod/bean (see page 7)
2 teaspoons ground ginger
2 balls stem ginger in syrup, finely chopped, plus extra to decorate
4 tablespoons syrup from the jar of stem ginger
the grated zest and freshly squeezed juice of 4 limes, plus extra to decorate

1 milkshake glass

Makes 400 ml/14 oz. and serves 1

Begin by making the Ginger and Lime Ice Cream or use a good-quality store-bought ice cream. Follow the instructions for making Vanilla Ice Cream on page 7, omitting the vanilla pod/bean and whisking the ground ginger into the cream in the saucepan instead. Add the chopped stem ginger, ginger syrup and the zest and juice of the limes to the custard before chilling.

Place the glasses in the freezer to chill for a few minutes.

For the shake, place a scoop of ice cream at the bottom of the glass before adding the rum and half of the ginger beer. Once the foam stops rising, add another scoop of ice cream before adding the remaining ginger beer. Top with an extra scoop of ice cream if you want to, and decorate with chopped stem ginger and lime zest.

I adore the nutty sweetness of Frangelico and, when paired with the irresistible additions of hazelnut gelato and Nutella, this two-toned hard shake is as classy as it gets.

Hazelnut Martini

50 ml/1 3/4 oz. ice-cold vodka
30 ml/1 oz. Frangelico
125 ml/1/2 cup whole milk
45 ml/3 tablespoons double/heavy cream
4 scoops Hazelnut Ice Cream (see below)
½ teaspoon xanthan gum (optional)
1 generous tablespoon chocolate hazelnut spread, such as Nutella

Hazelnut Ice Cream
1 quantity Vanilla Ice Cream (see page 7)
125 g/1/3 cup unsweetened hazelnut purée/paste
a pinch of salt

5 martini glasses

Makes 500 ml/17 oz. and serves 5

Begin by making the Hazelnut Ice Cream or use a good-quality store-bought ice cream. Follow the instructions for making Vanilla Ice Cream on page 7, but add the hazelnut purée/paste and a small pinch of salt to the hot custard and whisk them in before chilling.

Place the glasses in the freezer to chill for a few minutes.

For the shake, put the vodka, Frangelico, milk, cream and ice cream in a blender and pulse until smooth. Add the xanthan gum, if using (it will help to thicken the shake without affecting the flavour), and blend thoroughly.

Divide half of the milkshake between your chilled glasses (it should only go halfway up the glasses). Add the Nutella to the remaining milkshake and blend until smooth.

Tilt the martini glasses gently and carefully pour the Nutella milkshake on top. This will create a pretty two-tone, layered effect.

Shake, Rattle & Roll

A chocolate malt shake spiked with Bourbon, my Harlem Shake is sure to get you shuffling around on the dance floor.

Harlem Shake

8 scoops Vanilla Ice Cream (see page 7)
4 tablespoons malted milk powder
4 tablespoons Chocolate Fudge Sauce (see page 8), plus 2 tablespoons to serve
120 ml/4 oz. bourbon
canned whipped cream
a handful of mini marshmallows
grated dark/bittersweet chocolate

a handheld electric blender or milkshake mixer

2 milkshake glasses

Makes 800 ml/27 oz. and serves 2

Place the glasses in the freezer to chill for a few minutes.

Put the ice cream, malted milk powder, Chocolate Fudge Sauce and the bourbon into a jug/pitcher. Pulse lightly with a handheld electric blender, then switch to blending by hand, to help keep your shake nice and thick. If you have a milkshake mixer, blend it in the cup of the mixer until smooth and thick.

Divide the shake between two milkshake glasses and top with a squirt of canned whipped cream, some mini marshmallows and grated chocolate.

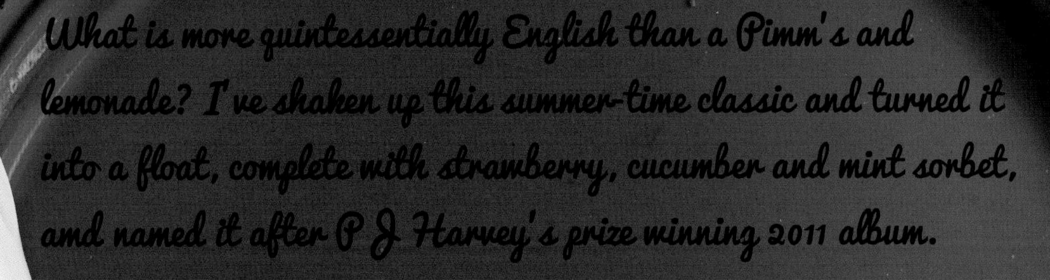

What is more quintessentially English than a Pimm's and lemonade? I've shaken up this summer-time classic and turned it into a float, complete with strawberry, cucumber and mint sorbet, amd named it after P J Harvey's prize winning 2011 album.

Let England Shake

300 ml/1¼ cups fizzy lemonade
100 ml/3½ oz. summer fruit cup, such as Pimm's
4 scoops Strawberry Sorbet (see page 7)
2 sprigs of fresh mint, to serve

2 highball tumblers

Makes 600 ml/20 oz. and serves 2

Place the glasses in the freezer to chill for a few minutes.

Mix the lemonade and Pimm's together in a jug/pitcher. Put 2 scoops of sorbet into each glass and top up with the Pimm's and lemonade. It will fizz up and the more you stir it the more quickly the sorbet will melt. Garnish each glass with a sprig of mint and serve, preferably in a pretty garden in early summer.

Based on Elvis' favourite fried banana and peanut butter sandwich, you can't help falling in love with this king of hard shakes. So, if you're hungry tonight, The Elvis will be always on your mind.

The Elvis

4 scoops Peanut Butter Ice Cream (see below)
80 ml/2¾ oz. crème de banane, plus extra for the glasses
30 ml/1 oz. vanilla vodka
1 large banana, peeled and roughly chopped
100 ml/⅓ cup whole milk
60 g/½ cup chocolate shavings, plus extra to serve
canned whipped cream

Peanut Butter Ice Cream
1 quantity of Vanilla Ice Cream (see page 7)
125 g/generous ½ cup smooth peanut butter
a pinch of salt

2 milkshake glasses

Makes 600 ml/20 oz. and serves 2

Begin by making the Peanut Butter Ice Cream or use a good-quality store-bought ice cream. Follow the instructions for making Vanilla Ice Cream on page 7, but add the peanut butter and a small pinch of salt to the hot custard and whisk in thoroughly before chilling.

Place the glasses in the freezer to chill for a few minutes. Once chilled, decorate the glasses with chocolate. Place the shavings in a shallow bowl wide enough to fit the glasses. Coat the rim of each glass with crème de banane before upturning them into the chocolate. Twist to coat the rims, then set aside.

Place all the ingredients except for the cream and extra chocolate shavings in a blender and pulse until smooth and thick. Divide between the glasses and top with a squirt of cream and a sprinkling of chocolate shavings. Thank you very much!

ELVIS

You won't need to be standing alone to enjoy this boozy blueberry shake. After trying one of these Violet Beauregarde-hued drinks, there'll be a dream in your heart to find a refill as quickly as humanly possible.

Blue Moon

2 tablespoons Blueberry Syrup (see page 9)
200 ml/7 oz. Blue Curaçao
100 ml/3½ oz. whole milk
2 scoops Blueberry Ice Cream (see below)
canned whipped cream
a handful of fresh blueberries, to garnish

Blueberry Ice Cream
200 g/1½ cup fresh or frozen blueberries
2 tablespoons caster/granulated sugar
1 quantity Vanilla Ice Cream made without the vanilla pod/bean (see page 7)

2 milkshake glasses

Makes 400 ml/14 oz. and serves 2

Begin by making the Blueberry Ice Cream. Put the blueberries, sugar and 2 tablespoons of water in a saucepan or pot set over a gentle heat. Stir until the sugar has dissolved, then increase the heat until the blueberries have completely broken down. Pass the blueberries through a fine mesh sieve/strainer and discard the seeds and skins, reserving the juice. Follow the instructions for making Vanilla Ice Cream on page 7, omitting the vanilla pod/bean. Once you have poured the custard into a cold jug/pitcher, stir in the blueberry juice until the mix is streak-free.

Place the glasses in the freezer to chill for a few minutes.

Spoon 1 tablespoon of blueberry syrup in the base of each glass. Blend the Blue Curaçao, milk and ice cream together until smooth and thick. Divide the milkshake between the glasses. Top with a squirt of cream and blueberries.

What is a St Patrick's Day celebration without a generous glug of hard liquor? Party in good Irish style with this gloriously green mint and whiskey hard shake.

Three Sheets to the Wind Shamrock Shake

10 scoops Peppermint Ice Cream (see below)
60 ml/2 oz. crème de menthe
120 ml/4 oz. whiskey
100 ml/⅓ cup whole milk

Peppermint Ice Cream
1 quantity Vanilla Ice Cream made without the vanilla pod/bean (see page 7)
1 teaspoon pure peppermint extract
2 teaspoons green food colouring

To Serve
canned whipped cream
3 maraschino cherries
fresh mint sprigs

a handheld electric blender or milkshake mixer

3 milkshake glasses

Makes 825 ml/28 oz. and serves 3

Begin by making the Peppermint Ice Cream or use a good-quality store-bought ice cream. Follow the instructions for making Vanilla Ice Cream on page 7, omitting the vanilla pod/bean. Once the custard has been made, stir in the peppermint extract and green food colouring before chilling.

Place the glasses in the freezer to chill for a few minutes.

Put the ice cream, crème de menthe, whiskey and milk in a jug/pitcher. Pulse lightly with a handheld electric blender, then switch to blending by hand, to help keep your shake nice and thick. If you have a milkshake mixer, blend it in the cup of the mixer until smooth and thick.

Divide the shake between the glasses, leaving plenty of room for whipped cream.

Top each with a generous squirt of cream, a maraschino cherry and a sprig of mint.

Dance along to the Piña Colada Song, while slurping from hollowed out pineapples filled with these kitsch cocktail shakes. Fresh, sweet and tropical, you won't even need to get caught in the rain to enjoy them!

Piña Colada

2 small pineapples
1 tablespoon coconut oil
4 scoops Coconut Ice Cream (see below)
200 ml/¾ cup coconut cream
150 ml/5 oz. golden rum
50 ml/1¾ oz. coconut liqueur, such as Malibu
maraschino cherries, to serve

Coconut Ice Cream
1 quantity Vanilla Ice Cream made without the vanilla pod/bean (see page 7)
50 ml/1¾ oz. coconut liqueur, such as Malibu
175 ml/⅔ cup coconut cream

a pineapple corer

Makes 800 ml/ 28 oz. and serves 2

Begin by making the Coconut Ice Cream or use a good-quality store-bought ice cream. Follow the instructions for making Vanilla Ice Cream on page 7, omitting the vanilla pod/bean and swap half of the cream for coconut cream and add the coconut liqueur.

Slice the top off each pineapple and reserve to make lids. Use a pineapple corer to remove the flesh of each fruit and pop your pineapple cups in the fridge until needed.

Place the pineapple flesh in a food processor and liquidize. Push the pineapple purée through a fine mesh sieve/strainer and discard any pulp. Blend 250 ml/1 cup of the resulting pineapple juice with the coconut oil, coconut ice cream, coconut cream, rum and coconut liqueur together, until thick and smooth.

Divide the milkshake between the chilled pineapple cups. Garnish the edge of each pineapple with 2 maraschino cherries pushed onto a cocktail umbrella.

This shake is Christmas in a glass, by which I mean a hint of spice and all the booze you can chuck at it.

Jingle Bell Rock

Begin by making the Honey Ice Cream or use a good-quality store-bought ice cream. Follow the instructions for making Vanilla Ice Cream on page 7, omitting the vanilla pod/bean and replacing the sugar with the honey.

Place the glasses in the freezer to chill for a few minutes.

Stir the whisky, pimento dram/allspice liqueur, Angostura bitters and lemon juice together, before blending through the ice cream, followed by the milk. Divide the milkshake between the glasses and dust with a grating of fresh nutmeg. Top each with a vanilla-soaked maraschino cherry – the cherries might sink, but it will be a nice surprise to discover after you've finished slurping.

200 ml/7 oz. whisky
50 ml/$1\frac{3}{4}$ oz. pimento dram/allspice liqueur
1 tablespoon Angostura bitters
freshly squeezed juice of $\frac{1}{2}$ lemon
3 scoops Honey Ice Cream (see below)
100 ml/$\frac{1}{3}$ cup whole milk

Honey Ice Cream
1 quantity Vanilla Ice Cream made without the vanilla pod/bean or sugar (see page 7)
50 g/3 tablespoons runny honey

To serve
grated fresh nutmeg
vanilla vodka-soaked maraschino cherries

6 lowball tumblers

Makes 800 ml/28 oz. and serves 6

Index

CLASSI